ECOLOGY
in Your Everyday Life

REAL WORLD SCIENCE

E Enslow Publishing
101 W. 23rd Street
Suite 240
New York, NY 10011
USA
enslow.com

Lisa Idzikowski

Published in 2020 by Enslow Publishing, LLC
101 W. 23rd Street, Suite 240, New York, NY 10011

Copyright © 2020 by Enslow Publishing, LLC.

All rights reserved.

No part of this book may be reproduced by any means without the written permission of the publisher.

Library of Congress Cataloging-in-Publication Data

Names: Idzikowski, Lisa, author.
Title: Ecology in your everyday life / Lisa Idzikowski.
Description: New York: Enslow Publishing, 2020. | Series: Real world science | Includes bibliographical references and index. | Audience: Grade 5 to 8.
Identifiers: LCCN 2018051336| ISBN 9781978507647 (library bound) | ISBN 9781978509498 (pbk.)
Subjects: LCSH: Ecology–Juvenile literature.
Classification: LCC QH541.14 .I35 j2020 | DDC 577–dc23
LC record available at https://lccn.loc.gov/2018051336

Printed in the United States of America

To Our Readers: We have done our best to make sure all website addresses in this book were active and appropriate when we went to press. However, the author and the publisher have no control over and assume no liability for the material available on those websites or on any websites they may link to. Any comments or suggestions can be sent by email to customerservice@enslow.com.

Photo Credits: Cover, p. 1 Marian Weyo/ Shutterstock.com; cover, p. 1 (science icons), back cover pattern kotoffei/Shutterstock.com; cover, p. 1 (globe graphic) Elkersh/Shutterstock.com; cover, interior pages (circular pattern) John_Dakapu/Shutterstock.com; p. 5 EpicStockMedia/Shutterstock.com; p. 6 JVrublevskaya/Shutterstock.com; p. 8 Sakurra/Shutterstock.com; p. 10 U.S. Geological Survey/Getty Images; p. 12 Peter Hermes Furian/Shutterstock.com; p. 14 Santhosh Varghese/Shutterstock.com; p. 18 Rainer Lesniewski/Shutterstock.com; p. 20 (top) milosk50/Shutterstock.com; p. 20 (bottom) Josef Hanus/Shutterstock.com; p. 23 Harvepino/Shutterstock.com; p. 24 Multiverse/Shutterstock.com; pp. 29, 52 Designua/Shutterstock.com; p. 31 Minerva Studio/Shutterstock.com; p. 33 Chz_mhOng/Shutterstock.com; p. 35 Gina Hagler Shutterstock.com; p. 40 BlueRingMedia/Shutterstock.com; p. 41 mapichai/Shutterstock.com; p. 44 Frances Roberts/Alamy Stock Photo; p. 48 Nerdist72/Shutterstock.com; p. 50 VectorMine/Shutterstock.com.

Contents

Introduction 4

■ **Chapter 1**
Nature Is All Around 7

■ **Chapter 2**
Matter and Energy—It's a Big Deal 17

■ **Chapter 3**
Where We Live 28

■ **Chapter 4**
Resources Are Crucial 38

■ **Chapter 5**
Environmental Issues Affect Us All 47

Chapter Notes 58
Glossary 61
Further Reading 62
Index 63

Introduction

Consider what may be happening around the world at any given moment. Kids rush to school on buses. Adults drive to work in vehicles. Bees pollinate flowers. Worms chew up bits of soil. Backpackers hike forested trails and watch for local wildlife. Farmers plow fields and keep livestock. Pelicans spear fish. Whales gulp krill. Indigenous or First Nations people gather and hunt for food. Families plant seeds or harvest vegetables. Mushrooms and bacteria deal with decaying matter. In some way, in some place, people interact with other people and the environment. They may interact with animals and plants there, too. Animals interact with each other, with plants, and perhaps people in their surroundings. All living things interact with each other and the nonliving. Ecologists are the scientists that study these things.

Ecology is a science. It tries to understand all the many interactions that occur each second of every day. People inhabit almost every corner of the globe. They share that space with about 1.9 million other animal species.[1] Many people live in city communities and might think they don't have much contact with ecological systems. Consider this: about one million species of insects have been identified and named.[2] Who hasn't been bitten by a mosquito or stung by a bee? Who hasn't found a silverfish, cockroach, or beetle in their home at one time or another? The number of bird species is much smaller. Still, who hasn't looked up at the sky to see a sparrow, starling, pigeon, or crow swooping through the air? Think about health. What if someone is sick and needs to take medicine? Many medicines have been developed from plant sources. What about eating? Most people enjoy a

Introduction

A late summer garden in the backyard is one example of how people interact with the environment. Ripe vegetables, colorful flowers, and tangy herbs reward the gardener after a successful season of planting and growing.

variety of food. Everything people eat can be traced back to a green plant. Plants are the only ones that produce food for every living thing on the planet. Then there's beauty. Who hasn't enjoyed seeing a rosy sunset, snow-capped mountain, or autumn-colored forest?

Ecology in Your Everyday Life

At one time, people probably thought forests had only a few purposes. They provided wood for fires or home building. Trees had to be cleared out to make way for farms and agriculture. People might have believed that animals were for food alone, either to be hunted or raised as livestock. What about the oceans? Something so vast could never be emptied of fish or polluted with trash. Of course, many individuals have helped the world understand the workings of nature and the effects of misunderstanding. Cut down too many forests, and erosion or air pollution may result. Overhunting can cause extinction or near extinction. No one ever thought that the passenger pigeon or American bison could disappear from the planet.

How about things people do every day almost without thinking? Kids help keep things out of the trash. They recycle paper, plastic, and glass. Other students use water containers instead of plastic bottles. Some families have compost heaps and turn vegetable scraps into dirt. Other individuals outfit their homes with solar panels to save energy. Think about it. Ecology is important to everyone on Earth. It is an interesting part of everyday life!

Nature Is All Around

Chapter 1

Look outside and around the neighborhood. Bike down the street or skateboard to a nearby park. Ride a bus to school or drive to the grocery store. Are towering trees or scrubby grasses part of the scene? Do fields of wheat ripple in the breeze? Are needle-covered trees standing tall? Or do prickly cacti surround the setting? People live in all sorts of places. Different types of plants and animals live there, too. All the members of a specific plant or animal species that live together in one area make up a population. This includes people. All the populations in one place comprise a community.[1]

Imagine a community near New York City. London plane trees or Norway maples shade sidewalks. Eastern gray squirrels chase each other up and down tree trunks.[2] Bald eagles soar. Monarch butterflies float among flowers.[3] Picture another community near Boulder, Colorado. Dragonflies zip through the air, and prairie dogs bark from their dens.[4] In all communities, there are biotic, or living, things. But that's not all. What about soil, rocks, water, sunlight, and oxygen? These abiotic, or nonliving, features are vital for communities, too. Without them, living beings cannot survive.

Natural Systems Are Always Changing

Changes happen all the time. Changes occur in individual organisms and entire ecosystems. They also happen to populations and communities. Increases and decreases in the number of plants or animals in a population or community are common. There are several reasons for this. Living beings interact with

Ecology in Your Everyday Life

This picturesque street in Greenwich Village shows how even in a large city such as New York, plant and animal life is not far away. Nearby parks or natural areas are communities filled with a variety of living things.

each other and nonliving elements in their surroundings. All these interactions within natural systems produce change.

Many animals, such as squirrels, normally give birth and raise young every year. So the number of squirrels in that population will go up. During the year, some will die. This causes the population of squirrels to go down. Typically, the number of individuals in a population remains pretty much the same.

Nature Is All Around

Let's say one year, conditions are different. One group of squirrels has an overabundant supply of food. Or perhaps they have found a perfect living space. Maybe the weather is ideal. All this could help them reproduce faster. Then their numbers might increase quicker than normal. Or the opposite could occur. Maybe many squirrels die from a rare disease. Or bad weather causes their food supply to be low. Some don't have enough to eat and starve. Perhaps where they live is destroyed by a natural disaster or lost to housing development. What if there are more predators eating this group of squirrels? In this situation, their numbers might decrease quicker than normal. Resources such

Saving an American Symbol

Flying high with its white head and tail feathers is the American bald eagle. At one time, the national bird was on the road to extinction. By 1940, Congress passed the Bald Eagle Protection Act to help eagle populations. It didn't work. A loss of habitat, shooting, and the chemical DDT further reduced eagle numbers. There were only 487 nesting pairs left. It looked like bald eagles would vanish. In 1972, the Environmental Protection Agency banned DDT use in the United States. In a few years, the bald eagle was listed as an endangered species. Then it could be protected under the Endangered Species Act. Because of this and other conservation efforts, bald eagle populations recovered. It is an ecological success story.

Ecology in Your Everyday Life

Eastern grey squirrels are common rodents living in the eastern United States. They prefer living in forested areas where they eat and travel in the canopy to avoid predators.

as food, water, living space, and weather have a large impact on organisms. These limiting factors matter.

Animals can move to different areas. Like kids moving into or out of a school, animals can move into or out of a population. Dragonflies, for instance, may fly into a new area to live. Around two years of age, beavers leave their home to find a new place for themselves.[5]

Many Interactions in Natural Systems

Like brothers and sisters in a household, or kids in a classroom, plants and animals are always interacting. Imagine being really hungry one night at dinner. Everyone has eaten, and one slice of pizza is left. Do you eat it so no one else does, or do you share it? Sometimes, animals and plants make similar decisions.

Suppose there are two animal populations living among the oak trees of a city park. Both squirrels and chipmunks love to eat acorns. Both depend on this food source. Most years, there would be enough acorns for all the animals to eat. Squirrels and chipmunks would be healthy and reproduce normally. Both would have acorns stored for the winter. What if there was a shortage one year?[6] Squirrels and chipmunks would want the same food. They would compete for the available acorns. Some would go hungry. Some might die as a result. Competition occurs when different organisms struggle for the same resources—in this case, food.

Squirrels are not the only organisms to profit in this situation. The oak trees benefit, too! Sure, the squirrels eat a lot of acorns, but they also save some for winter when food is scarce. To do this, they bury acorns in the ground. They don't always remember where their loot is buried, so some acorns are never eaten and remain buried. When conditions are right, the acorns, which are oak seeds, start to grow.[7] In this way, the next generation of oak

Ecology in Your Everyday Life

From its perch, an adult Cooper's hawk surveys the area possibly searching for its next meal. With a long tail and short wings it maneuvers through woodlands.

trees begins its life. Plant and animal benefit. Squirrels get food, and oak trees have their seeds scattered to grow. They are in a relationship of mutualism.

Another type of interaction can occur in this squirrel, chipmunk, and oak tree ecosystem. It involves a predator, the Cooper's hawk. These birds live in forests and woodlands. They have learned to live in city neighborhoods, parks, and even backyards. Cooper's hawks nest in trees, and they eat many small birds and mammals. The hawks prey on squirrels and chipmunks, too.[8] Hawks, squirrels, and chipmunks interact in a relationship of predation.

Activity: Can Animals Avoid Being Eaten by Predators?

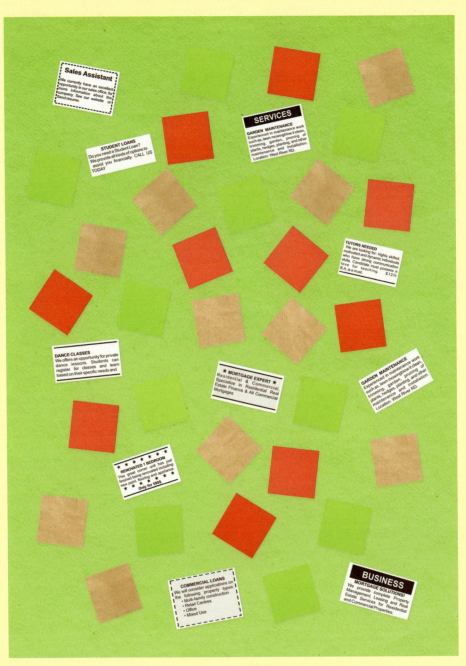

Nature Is All Around

All animals need food to survive. Animals cannot make their own food, so they eat other creatures. Is there a way for animals to avoid being eaten by predators? Yes. Many use a variety of ways to stay alive. Some fight their attackers, some run away, some hide, and some use camouflage to avoid being seen.

Things You Will Need:

- a pencil
- paper
- 2 sheets of newspaper (mostly black-and-white "want ads")
- 2 sheets of green paper
- 1 sheet of red paper
- 1 sheet of brown paper
- scissors
- a helper (friend, sibling, or adult)
- a clock or timer

■ **1.** Have the pencil and paper ready to record observations.

■ **2.** Set out two sheets of paper—one newsprint, one green sheet.

■ **3.** Cut ten ½-inch squares of each: newsprint, green, red, and brown paper. You will have a total of forty squares.

■ **4.** Explain to your helper that they are going to pretend to be a bird that is hunting for caterpillars to eat. They will try and capture as many of the ½-inch square paper "caterpillars" as they can at a time.

■ **5.** Ask the helper to walk a short distance away.

Ecology in Your Everyday Life

■ **6.** Place all the paper "caterpillars" on the newsprint sheet.

■ **7.** Get ready to use the timer. Have the helper come and capture as many caterpillars as they can in twenty seconds. Record the number of captured caterpillars. Be sure to record whether they are newsprint, green, red, or brown.

■ **8.** Repeat steps 6 and 7 four more times.

■ **9.** Now repeat steps 6 through 8 using the green sheet of paper in place of the newsprint paper.

■ **10.** After finishing, answer these questions: What color caterpillar was captured the most with the newsprint as background? What color caterpillar was captured the most with the green paper as background? How is this like camouflage in nature?

Matter and Energy—It's a Big Deal

Chapter 2

It's late afternoon on a Friday. Kids are scrambling to finish homework. They are hungry, and they are thinking of their early evening soccer game. Many families are wondering what's for dinner. Some parents are home early and preparing meals. At one house, they are cooking pork chops, broccoli, carrots, and mashed potatoes smothered in mushroom gravy. Sounds delicious, doesn't it?

What about the soccer game? Forwards are dreaming about scoring the perfect goal. Midfielders are planning their best dribbling and passing moves. Defenders are getting ready to kick strong and far. Goalies are wondering if they will be able to grab a spinning ball or stop a goal.

Interestingly, the dinner and the game can be comparable to cycles of energy and matter in nature. Each has individual parts that work together. At dinner, protein and vegetables form a meal. At the game, individual players in various positions make up a team, working toward the common goal of beating their opponents. When the parts of the meal or the team function well, the elements come together and form a system, or a whole, like puzzle pieces that fit together. This happens in nature, too.

Ecosystem Energy Cycles

Just as people eat food to gain energy, so do animals. Our imaginary family enjoyed a meal of pork chops—the meat from another animal. They ate vegetables, which are plants. In the wild, most animals are forever searching for food. Like people, they can't make their own food in their bodies. People and animals

Ecology in Your Everyday Life

must eat other living organisms. Plants, algae, and some bacteria are the only living things that can make food. These producers make food using sunlight, water, and carbon dioxide in a process called photosynthesis. This food takes the form of sugars and starches and is stored in the plant. People and animals—consumers—eat the plants. Or they eat other animals that have already eaten plants.[1] This is the only way that people and animals have food. Energy that is obtained from the food eaten is vital.

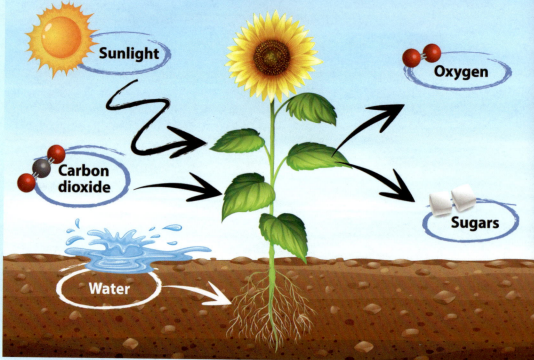

Through photosynthesis, plants such as this sunflower use the energy of the sun, carbon dioxide, and water to produce sugars and oxygen.

Matter and Energy—It's a Big Deal

Without it, animals and people would not be able to grow, move, reproduce, and carry on with life.

Imagine being in a backyard or garden. This ecosystem may have grass, flowers, and trees. Depending on the location, different types of animals live there, too, including birds, insects, reptiles, amphibians, and mammals. In some gardens, people plant native flowers to attract insects. One favorite is the milkweed. As a green plant, it is a producer. Insects use it as a food source. They are its consumers. In fact, one insect depends on milkweed. The caterpillars of monarch butterflies eat no other plant. If there are no milkweed plants, female monarchs have nowhere to lay eggs. If that happens, there would be no new generation of monarch butterflies.[2] The cycle of energy doesn't end there. Backyards and gardens provide habitats for hungry birds looking for food. Caterpillars are a favorite food of birds. Robins are backyard-loving birds that readily gobble up caterpillars, including those of the monarch.

Mammals eat garden plants. Woodchucks or groundhogs chow down on grasses, berries, flowers, and tree bark.[3] They eat grasshoppers and other large insects. Because they eat plants and animals, these consumers are omnivores. What else lurks? The American toad hops around in most areas of eastern Canada and the United States. It likes moist spots where it snaps up insects, spiders, worms, and slugs.[4] Because the American toad eats other animals, it is a carnivore. Milkweed plants are eaten by monarchs, which are then eaten by robins. Grass is eaten by grasshoppers, which in turn are eaten by woodchucks. The producer makes food, and the consumer eats the producer—this is a simple food chain. In truth, many animals eat many things. The flow of food energy in an ecosystem is really a combination of different food chains that make up a food web.[5]

Ecology in Your Everyday Life

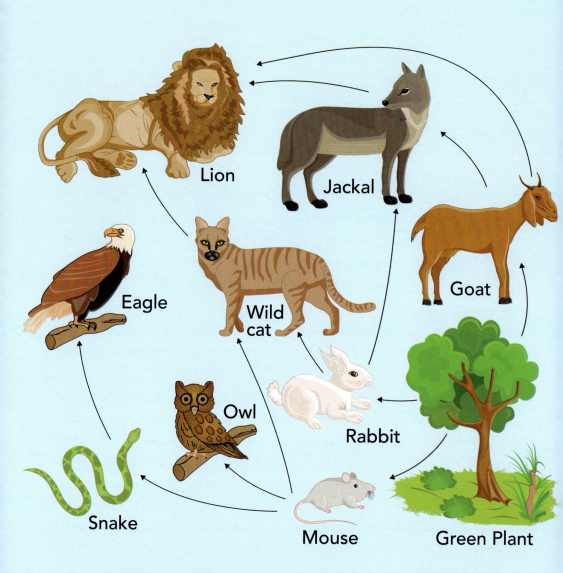

The flow of food energy in a food chain begins with green plants. And in an ecosystem, a variety of food chains interact to form a food web.

Matter and Energy—It's a Big Deal

Extreme Producers

Not all producers use energy from the sun to make food through photosynthesis. Some bacteria use energy from chemical reactions instead. These bacteria exist in extreme environments. Some live in volcanoes. Others are found in the hot springs of Yellowstone National Park. Additional varieties populate the deep ocean, either near cracks or on the seafloor. In these places, a numerous supply of certain chemicals occurs. This process of chemically producing food is called chemosynthesis.[6]

One of the most important elements of our imaginary meal is the mushroom. In nature, food is being produced and eaten, and organisms are living. They are also dying and producing waste. What happens to the waste and dead remains? Where do they go? A group of organisms—decomposers—deal with them. Earthworms, mushrooms, mold, mildew, and bacteria are common decomposers. They break down waste and dead organisms into chemicals that act as nutrients for living things. In this way, chemicals of the once living are recycled back into nature, available for use in the future.

Cycles of Matter in Ecosystems

Soccer balls and shin guards, uniforms and cleats. Pots and pans, cooking utensils, stoves and refrigerators. All things are made of matter. We live in it, around it, and with it. The same is true for the world of nature. Everything is composed of matter, but certain

Ecology in Your Everyday Life

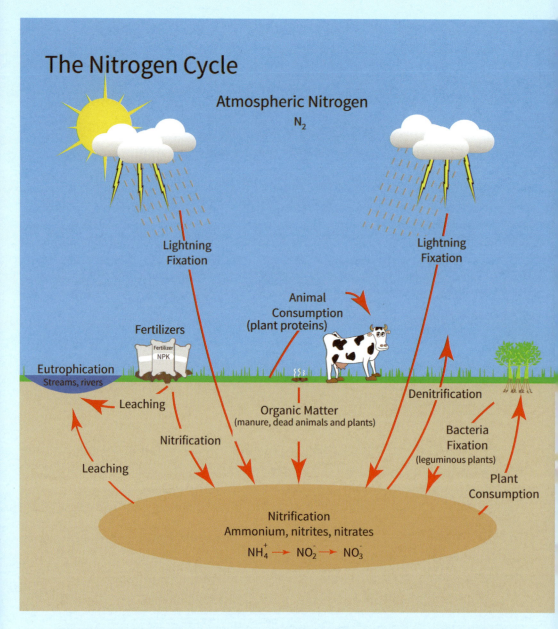

Nitrogen is an important chemical element in nature. It is available to natural systems because of a recycling process in nature, the nitrogen cycle.

chemicals are the most important in this system. They are water, oxygen, carbon, and nitrogen. These important substances cycle continuously through ecosystems. Without them, life on Earth would be impossible.

In backyards and gardens, like other habitats, water is a necessity. Depending on the region, different amounts of precipitation fall throughout the year. Soils get help to hold onto water. Earthworms shred, mix, chomp, and chew through the soil. All this action makes it easier for soil to keep the right amount of moisture.[7] The water cycle works as a giant recycling system. Water moves from one location to the next. Because of this, it is available for living things. Water evaporates and moves into the atmosphere as a gas. Up high, temperatures are cooler. The gas condenses and forms clouds. With time, heavy clouds release the water and it falls back to Earth as some type of precipitation, which then supports life.

Carbon and oxygen are team players. As part of the carbon-oxygen cycle, the two are in a push-pull relationship. Plants, such as grasses, take in carbon that animals, such as squirrels, breathe out as carbon dioxide. In turn, plants use the chemical along with energy from the sun for photosynthesis. Without the chemical powerhouse production of photosynthesis, animals—including people—would have no food. Oxygen is an added bonus of the process, forming an important part of the air that people and animals breathe.

Nitrogen is another important component in nature. Plants need it to grow and thrive. So do animals. It's true that air is in large part nitrogen. This airborne type is unusable. Instead, clumps of bacteria living on plant roots transform this form into one that is usable by plants. In grasslands, white prairie, purple prairie, and round-headed bush clover have the special bacteria.[8] In this win-win situation, bacteria feed on the plant and produce nitrogen for

Ecology in Your Everyday Life

the plant. Animals win, too. Herbivores, such as bison, elk, and deer, eat prairie plants and get the nitrogen that they need to stay fit and healthy. At some point, decomposers break down the nitrogen completely, and it flows back into the air to start this cycle again.

Activity: Does Water Speed Decomposition?

Molds are one organism that causes decomposition. Mold spores can be found everywhere. As decomposers, they play an important role in all ecosystems. Does decomposition occur at the same rate in all cases? Or are there ways to increase the rate at which decomposers work?

Things You Will Need:

- a pencil
- paper
- 2 one-gallon Ziploc bags
- masking tape
- a marker
- 1 medium-sized potato
- a butter knife
- a measuring cup
- ½ cup of soil

1. Have the pencil and paper ready to record observations.
2. Take out the two gallon-sized plastic bags.
3. Place a piece of masking tape labeled "dry" on one bag.
4. Place a piece of masking tape labeled "wet" on the other bag.
5. Cut the potato in half, and place a piece in each labeled bag.
6. Measure out ¼ cup of soil and place it in one bag. Do the same with the other bag.
7. Moisten the soil in the bag labeled "wet."

Ecology in Your Everyday Life

A decomposing potato is a disgusting sight—and a fascinating look at ecology at work in your own home.

26

Matter and Energy—It's a Big Deal

■ **8.** Seal both plastic bags, and place them side-by-side on the counter. Do not open the bags.

■ **9.** Check the bags twice a day for two weeks. Record observations.

■ **10.** Answer these questions: What happened to the potatoes? Did the same happen to both the wet and dry potatoes at the same time? Why or why not? Explain what this would look like in nature.

■ **11.** When finished, throw away the bags, potatoes, and dirt.

Chapter 3
Where We Live

Can anyone say no to their favorite foods? Would people give up corn chips or cereal? How about french fries or potato chips? Could they get along without bread for peanut butter and jelly sandwiches? What about tortillas for tacos? And pizza. Without a crust layered with cheese, sauce, and pepperoni, it just wouldn't seem like pizza! Many of these foods have ingredients that in one way or another originally came from wild plants. Let's not forget about foods that are or come from animals. Many people around the world eat wild animals. An enormous amount of fish and other seafood are caught and consumed each year.[1]

How many people take aspirin for sore muscles, headaches, or fever? Ever had strep throat or an ear infection? Penicillin, a common drug, is prescribed to fight bacterial infections. It was discovered by accident. A scientist happened upon a special kind of mold. He found that the mold could keep bacteria from growing. Not long after, others figured out how to turn the mold into medicine. From then on, penicillin saved the lives of many people.[2] Many other medicines have been developed from wild plants. Thousands of plant species are used for medicine each year around the world.[3] In the United States, about one-fourth of all prescriptions contain elements that come from plants.[4]

The Web of Life

The variety of living organisms on Earth is enormous. Its plants and animals are all connected and form a richness of life. Biodiversity is the term used to describe this web of life. It includes genetic

Where We Live

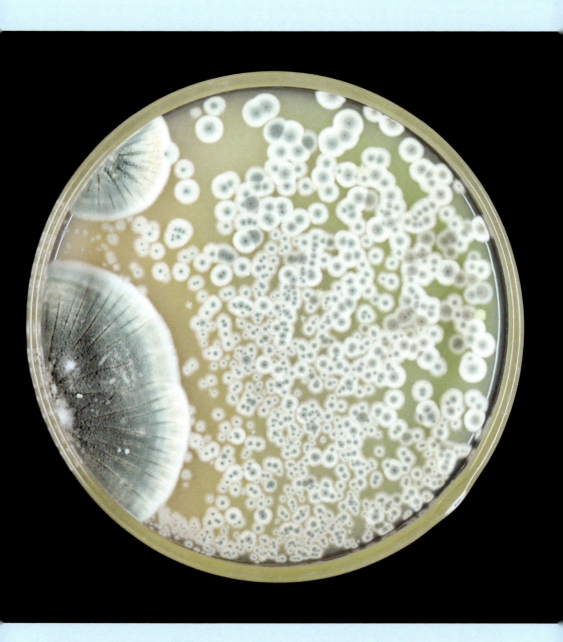

There are many different types of *Penicillium* mold. Some are harmful, while others are beneficial. *Penicillium* mold can be grown under lab conditions on prepared agar plates.

Ecology in Your Everyday Life

differences in species, as well as how they interact with one another. It also includes all habitats, communities, and ecosystems in the world. This amazing variety of living organisms is very special. Scientists emphasize that people can't live without it. Food and medicine may matter the most, but humans benefit in so many other ways. People have used plant products, such as cotton, for clothing. Wood from trees has provided materials for building and for keeping warm. Our clean air, drinkable water, and good agricultural soil happen because of ecosystems that work properly.[5] Ecologists make a strong case. People must work to protect and encourage the world's natural systems!

Biomes

Earth has a grand lineup of organisms. Where do they all live? Are they scattered about or close together? What is the temperature like—is it cold or warm? Is there land, water, or ice and snow? How do trees and other plants grow? Temperature, water, sunlight, wind, rocks, and soil are factors that affect natural areas. The workings of temperature, water, sunlight, and wind form weather. Weather in a place over a long time, or climate, determines where certain plants and animals exist. Because of this, the sorts of plants, animals, and climate in an area form ecosystems called biomes.

Across the globe, there are two main types of biomes. Many people are familiar with some that appear on land. Forests of several kinds are one example. They include tropical rain forests, where it is always warm, wet, and rainy; coniferous tree forests, which are usually cool or cold and where most of the trees have cones and needles; and deciduous tree forests that experience the four seasons of spring, summer, fall, and winter. Here, most of the trees lose their leaves in autumn. Desert, grassland, and tundra biomes also are found on land. In the desert, it is always

Where We Live

Red deer graze on grass in their forest home in Piedmont, Italy, in autumn. Deciduous forests are woodlands found in many parts of the world where trees such as oak, maple, ash, and beech lose their leaves during the fall season.

Ecology in Your Everyday Life

extremely dry. Grasslands are dry, too, but they get more rain than the desert. The tundra is not only very dry, but also especially cold.

Watery worlds make up the other main type of biome. This world is divided according to the water found there, either fresh or salt. Ponds, lakes, streams, rivers, and wetlands contain mostly fresh water. Many kinds of organisms live in these freshwater biomes. Fish, birds, insects, and amphibians are some of the common consumers. Algae living in the shallow water use sunlight and carry on photosynthesis. They are the producers in freshwater habitats. Estuaries where freshwater rivers meet the ocean are another watery ecosystem. The water there is a mix of salt and fresh. The plants and animals living there are adapted to the

Alien Species

Many people might be surprised that they have seen an alien species. It is a common bird over much of the United States—the European starling! Starlings aren't native birds. They were brought over from Europe in 1890. The original group of sixty birds blossomed into millions. Some people like starlings. Like most non-natives, they cause problems. At bird feeders, they eat seed meant for native birds. Their nesting habits are a big problem. They out-compete native birds. Woodpeckers, bluebirds, chickadees, and swallows have problems nesting because starlings will take over their nests. Sometimes, they even destroy other birds' eggs.[6]

Where We Live

A sea otter munches on sea urchin in Monterey Bay, California. Sea otters are weasel-like mammals that spend much of their time in the water.

mixture. Oceans have many different zones, or depths, of water. Many wonderful species call the different ocean zones their home. Oceans and estuaries are both considered saltwater biomes.

In biomes, certain animals play a role that keeps the ecosystem operating at its best. Many of the ecosystem's other species

Ecology in Your Everyday Life

wouldn't survive otherwise. These keystone species are a necessary part of land and water biomes. They can be found in ecosystems all around the world. In the grasslands of Africa, for instance, elephants are keystone species. Elephants either eat trees and small shrubs or knock them over. This action keeps grasses growing freely, which in turn supports other animals. Zebras and antelopes are grazers, eating the grass. They become food for predators, the lions and hyenas, which then have plenty of food to maintain their lives.

Another keystone species, the sea otter, lives in the Pacific Ocean ecosystem, along the coasts of North America and Asia.[7] One of the otter's favorite foods is sea urchin. By chowing down on urchins, sea otters keep their ecosystem healthy. If there aren't enough otters, the numbers of urchins go up. That causes the urchin's food of choice, the giant kelp plant, to be eaten down to nothing. If that happens, all the creatures that depend on kelp die. This is exactly what happened in the past. Sea otters were once hunted for their fur. They almost became extinct. Not unexpectedly, kelp disappeared, as did many other members of that ocean ecosystem. Luckily, scientists came to the rescue and sea otters became a protected species. Eventually, the kelp forests came back, as did that ocean community.[8]

Activity: Be an Ecosystem Detective

Mini or small ecosystems are everywhere. Climate affects ecosystems and the plants and animals living in an area. Can plants and animals be found in very small places? Are they all different? Are they the same? Why do some live in one area and not the other?

A small pond can provide a living area for many plants and animals.

Ecology in Your Everyday Life

Things You Will Need:

- a pencil
- paper
- crayons, colored pencils, or markers
- rope or string, 8 to 10 feet (2.4 to 3 meters) long
- a watch, cell phone, or some other device to keep track of time

■ **1.** Choose a good weather day, not too hot, cold, or windy.

■ **2.** Find two different mini ecosystems. HINT: It could be a big stone in a backyard, a fallen log at a nearby park, the shoreline at a beach or pond, or a patch of moss on the side of a tree.

■ **3.** Have the pencil and paper ready to record observations.

■ **4.** Have colored pencils, crayons, or markers to draw your observations.

■ **5.** Using the rope or string, form a circle around your mini ecosystem. (It doesn't have to be a perfect circle; it will be the edge of your investigation area.)

■ **6.** Now be a detective. Observe what you see in your marked-out ecosystem.

■ **7.** If you know the name of something, write it down. If not, make a drawing of it. Also draw what you see.

■ **8.** Observe and record your mini ecosystem for thirty minutes. Include any animals that fly over your area.

■ **9.** Repeat steps 3 through 8 for your other mini ecosystem.

Where We Live

■ **10.** When finished, compare your two mini ecosystems.

■ **11.** Answer these questions: What plants or animals were the same or different in each spot? Were the conditions the same or different? What caused the similarities or differences? Explain what this might look like if the ecosystem was larger.

Chapter 4
Resources Are Crucial

Recall a time in the woods on a bright sunny day. Trees sway in the brisk wind. The sun peeks out from behind a few cumulous clouds. It sounds like a perfect place. Sun, trees, and wind—people use these three natural resources. Within limits, these renewable resources are replaced by nature. At another time, people turn on the lights, computer, and television. They turn up the heat or air conditioner. They toast a bagel and warm coffee in the microwave. Out in the community, cars speed, buses pick up passengers, trains leave from the downtown station, and planes take off from the airport. It could be a typical morning in many communities. People use electricity and fuel daily. Coal, oil, and natural gas—fossil fuels—provide much of this electricity and fuel. Because they take millions of years to form, they are considered nonrenewable resources.

Energy, resources, and climate change are tough topics. Scientists and nonscientists believe the world must reduce its use of nonrenewable energy sources. They argue that fossil fuels are running out. They also believe that fossil fuels are causing damage to the environment and natural systems. Not everyone agrees, and that is a large obstacle. How can problems be solved when time and energy are spent on disagreements? Fortunately, some countries, companies, and citizens are forging ahead. They are tackling the controversies. They are experimenting and refining solar power, green technologies, and biofuels.

Resources Are Crucial

Renewable Energy

Solar power traps and uses the sun's natural energy. Early in the 1800s, a scientist discovered that electricity could be obtained from sunlight. Solar power has come a long way.[1] At one time, this technology was expensive. It has gotten much cheaper. Solar panels can be seen in many places. Many homes and businesses already have them on rooftops. Solar panels are made from solar cells that are grouped together. When sunlight hits special

Solar panels on the roof collect sunlight to power this house. Because this renewable energy source is getting much cheaper to install, it can be an affordable choice for homeowners.

Ecology in Your Everyday Life

materials inside these cells, energy is captured and turned into electricity.[2]

Solar energy is not the only renewable energy source. Research continues to improve the potential for wind and biomass technology. Some scientists say that wind can provide plenty of power for the planet. Windmills or turbines change wind energy into electrical power. Long, large blades on these machines turn in the wind. In windy areas, large wind farms or groups of windmills are placed. This energy source is promising. With wind energy, there is no greenhouse gas as with fossil fuels.[3]

Biomass or biofuel has different origins. Wood, algae, animal waste, and corn are a few common sources. Biomass produces energy in two different ways: heat is given off by burning it, or the material is made into fuel. A corn-gasoline fuel mix pumps into cars at many gas stations. Special types of corn are grown and tested in the Midwest for use in this biofuel. Does it stack up? Some argue that it's not worth it. Either too much energy is used to produce the fuel, or production emits unwanted greenhouse gases. Of course, there are scientists who want to keep looking into this fuel and its possible uses.

Look at that green scum on top of the water. It's algae. Earth's algae produce about half of the oxygen needed for the world's organisms. Besides that, these plants make energy-rich oil. Researchers, including those at NASA, have been getting on the algae bandwagon. Not only do algae make fuel, they also clean dirty water in the process![4] There is another big advantage: algae out-produce corn. Grown in the same size area, algae produce eighty times more oil than corn.

Wood has been used as fuel for a very long time. Now, researchers are finding ways to use leftover sawdust to produce biofuels. Researchers in Australia are testing fungus to see if they can help with this project.

Resources Are Crucial

In addition, cow waste has the potential to be an alternative fuel. Scientists are experimenting with fuel cells made with manure and other chemicals. Right now, it is producing small amounts of electricity. One can never predict the future, but this cheap source of matter may be powerful in the years to come.

Nonrenewable Resources

Coal, oil, and natural gas are the three major types of fossil fuels. Why the name "fossil fuel"? This is because these substances are formed from the remains of ancient living things—dead plants, animals, and other organisms from millions of years ago mixed with rock, sand, and mud. Year after year, layer after layer, the once living became sandwiched with the nonliving. The weight

What's in That Cell Phone?

A large amount of materials used by people in the United States is nonrenewable. These substances include minerals, metals, and products made from fossil fuels. Cell phone manufacturing relies on a variety of minerals. Over half of the parts of a mobile device are from mineral ores. Do you like to look at a nice, clear screen? Silica sand is the mineral source that gives a glass screen its clarity. What about the battery? Sphalerite is the mineral source that helps conduct electricity. You just can't wait to get the newest model? Think about all the nonrenewable materials that go into making a phone!

Ecology in Your Everyday Life

above pushed down with great pressure and heat. Slowly, the mixture changed, and over eons turned into nonrenewable fuel.

Coal is a solid fossil fuel. It is mined from within Earth, chopped apart, and hauled to the surface. It is plentiful in the United States. Electric power plants use coal to produce electricity. Oil is a liquid fossil fuel. It powers most things that people think about: cars, ships, motorcycles, planes, and trains. It also heats some homes. Many products, such as plastics, are made from oil or petroleum, too. Oil is found deep under the ground. After being pumped to the surface, oil must be refined to be of use. Natural gas is a

Coal is a dark brownish-black or black sedimentary rock. A big difference between coal and other rocks is that coal is formed from plant materials and not minerals.

Resources Are Crucial

gaseous fossil fuel. It often is found above oil in the same rocks. Natural gas is moved through pipelines to where it is needed. This fossil fuel is somewhat clean compared to oil and coal. It is used for many purposes in homes and businesses. One big problem is that it catches fire easily. Gas explosions have happened from leaky pipes.

There are problems with fossil fuels besides that of being nonrenewable. Debates rage on, but people the world over agree that fossil fuels have contributed to global warming. This is causing a number of effects. Average temperatures are increasing. Polar

Wildfires are an important part of certain natural ecosystems. The plants and animals living in those areas have evolved over time to thrive under those conditions.

Ecology in Your Everyday Life

ice and glaciers are melting. Sea levels are rising. Severe storms happen yearly. Weather patterns are changing. Devastating fires are sweeping over the land. Fire has been a usual event in ecosystems since time began. But with global warming, severe fires pop up more frequently. Yellowstone National Park is a good example. It has recovered from several horrible fires in roughly the last thirty years. These fires burned through an incredible amount of land. Forests, plants, and animals are adapted to survive fires. In fact, these ecosystems do better when fire sweeps the land every so often. The problem is that global warming is causing the conditions to be right for these giant fires to occur on a much more frequent basis, not the usual severe fires every one hundred to three hundred years as has been normal in the past.[5]

Activity: How Can Water Keep the Heat?

Solar power can be used to heat buildings. This method is called passive solar heating. These buildings are designed with large windows to let in the maximum amount of sunlight at times of the year when heat is needed. Special materials absorb the heat from the sun during the day. At night, the materials release heat to warm the space. Can different materials absorb more of the sun's heat?

Things You Will Need:

- a pencil
- paper
- scissors
- 3 sheets of construction paper: 1 white, 1 black, 1 medium blue
- 3 tall glass jars or cups that will hold ½ cup (118 mL) of water
- masking tape
- a measuring cup
- water
- a thermometer
- a watch, clock, or cell phone to keep track of time

* Do this on a day where bright sunlight from a window is available.

1. Have the pencil and paper ready to record observations.
2. Cut the construction paper to fit around each jar or cup.

Ecology in Your Everyday Life

■ **3.** Tape the paper to each container, making sure as much surface as possible is covered.

■ **4.** Fill each container with half a cup of water.

■ **5.** Take the temperature of each container. This is the starting temperature.

■ **6.** Place all three containers in a bright sunlit window.

■ **7.** Wait for 15 minutes, then take the water temperature of each container and record the data.

■ **8.** Repeat step 7 three more times.

■ **9.** Compare the temperatures recorded from steps 7 and 8 to the starting temperature.

■ **10.** Answer these questions: Did a difference in temperature occur? Which jar had the coolest temperature? Which had the warmest temperature? How would this matter to materials used in passive solar heating?

Environmental Issues Affect Us All

Chapter 5

The number of people on the planet continues to grow. By 2050, about 9.8 billion people will be living around the world. All these people will have an impact on the environment. It may be cutting down the forest, or it may be saving endangered species. It may be educating others about science, or it may be throwing away garbage that stuffs landfills or causes pollution. People will affect natural systems around the globe. These same natural systems will affect us all.

Water Pollution

It's laundry day. Individuals and families wash clothes every week, sometimes every day. What happens to the soapy water from washing machines? Where does it go? Laundry detergent used to contain a high amount of harmful chemicals. Phosphorus was the main culprit. It caused problems. Lakes clogged up, overgrown with algae. Fish and other organisms died. Thankfully, companies altered their detergents, which now cause less pollution. What about water from sinks, bathtubs, and toilets? In many places, it flows down the drain to treatment plants. There this sewage, or "dirty" water, makes its way through a series of treatments. After processing, it is much cleaner and ready to be used again, or returned to lakes, stream, and rivers.

What happens when supposedly safe drinking water is polluted? That is exactly what occurred in Flint, Michigan.[1] People were told that the brown water flowing from faucets was safe. It wasn't. Water treatment failed. People drank water filled with lead

Ecology in Your Everyday Life

Clean, fresh water is a necessity of life. Water treatment facilities use a variety of ways to treat water so that it is safe to use. But sometimes treatment fails and brown, polluted water flows out of faucets instead.

Environmental Issues Affect Us All

and iron. Some got sick, and others died. The state handed out bottled water. After more than two years of testing and changes to water systems, Flint's water tested clean and was declared safe to use again.

Imagine gorgeous gardens bursting with flowers, filled with ripe vegetables, and trimmed by the greenest grass. How do people achieve this feat? Many use fertilizers. Applied correctly and in small amounts, fertilizers are fine. Ramp that up to a grand scale, as in millions of acres of farmland, and watch out. Farmers ready their fields for planting in the spring. Many apply a mix of chemical fertilizers and animal manure to the land. Heavy rain is common. This combination causes chemicals to wash over the land as runoff. Eventually, the runoff reaches creeks, ponds, lakes, streams, and rivers. In the midwestern United States, this water flows into the Mississippi River. The mighty Mississippi travels south and empties into the ocean. What results is a "dead zone" in the Gulf of Mexico—a large area choked by algae where fish, shrimp, and an entire ecosystem of life disappears.[2]

Nature has a way to handle water pollution. Wetlands act as a natural water treatment plant. Water runs off land, enters a wetland, and is filtered. Sewage, manure, and some chemicals settle out of the water. In some instances, nature's way of cleaning has been as effective as a typical water treatment plant.[3] Keeping wetlands healthy and preventing them from being paved over for development is vital. Wetlands also stop flooding by absorbing water from extreme weather events.

Trashing the Earth

Let's talk trash. Statistics say every American produces more than 4 pounds (1.8 kilograms) of trash each day. Many young children learn the three R's—reduce, reuse, recycle—in school.

Ecology in Your Everyday Life

Landfills are as harmful to the environment as they are unsightly. Many communities have recycling centers, and homeowners are encouraged to recycle a variety of materials so that they don't end up in the trash.

Environmental Issues Affect Us All

Make a Difference!

There are easy ways to help the environment: Use reusable bags and containers. Refuse to use styrofoam of any kind. Turn off the lights, TV, computer, and other electronic devices when you are not using them. Avoid littering. Recycle. Take your bike or walk. Donate old clothes and sports equipment. It's not hard, so get started!

Are people putting this into action? Are they reducing what they use? Are they reusing or fixing things instead of tossing them out? Are they recycling? So much is thrown away. Most trash is either buried in landfills or burned. Burning or burying solid waste harms the environment in numerous ways. Air pollution, water contamination, ocean pollution, and habitat destruction happen. Think of all the simple ways to reduce trash. Don't buy and use water in disposable plastic bottles; buy a reusable water bottle. Don't go to the grocery store and have food packed in plastic bags; buy reusable bags and carry them wherever you go. Ocean ecosystems are dying, and tons of plastic garbage are the reason!

Deforestation Devastates Everyone

Decreasing pollution and trash is necessary. Natural areas must also be guarded and protected from demolition and development. Most of us live in houses, write on paper, and receive packaged items ordered over the internet. Some of us love to eat nuts such as pecans, almonds, cashews, or walnuts. Pancake and

Ecology in Your Everyday Life

waffle lovers smother their creations with maple syrup. Trees provide society with so many benefits, especially the oxygen that we all breathe. They also give people joy. Some people enjoy collecting colored leaves in the autumn or watching bats flit through branches while catching insects. Others like seeing

Gold mining has destroyed parts of this rain forest in the South American country of Guyana. About 30 percent of Earth is still covered in trees, but deforestation is slashing that amount yearly.

Environmental Issues Affect Us All

squirrels chasing each other down the trunks. Many people love decorating trees for the holidays.

All over the world, trees are in crisis. Deforestation, or the permanent destruction of forests, is increasing. Millions of acres have been destroyed. Half of tropical rain forests are gone. Problems are developing. As trees are cut or burned, they release greenhouse gases into the air. Not only are there fewer trees producing oxygen, but global warming is increased. Soil erodes without tree roots to hold it in place. The water cycle becomes unbalanced. More than half the water in the Amazon ecosystem is in its plants. When forests disappear, so do plant and animal species living there.[4]

Fewer plants create another downside—the loss of potential medical treatments. Beneficial medicines have been developed from forest trees and plants. Acne is an uncomfortable problem for some teens. Some medicines used to treat it contain a special chemical called salicin. Salicin was the basis for aspirin at one time. It originates from the bark of willow trees. Taxol is another valuable medicine. This lifesaving cancer drug came from the bark of yew trees.[5] It was discovered by scientists sent to sample tree bark in the forests of the northwestern United States. At first, the drug was made from harvested wild tree bark. Now, it is produced using chemicals and yew trees specifically grown for this reason. Researchers don't know where they'll find the next important ingredient for a new medicine. This is another reason to stop destroying forest ecosystems.

A Modern Mass Extinction Event

The extinction of animals and plants is a possibility whenever forests and other natural areas go under the bulldozer. Organisms have become extinct since the beginning of time, but now it is happening faster and faster. Scientists believe human actions

Ecology in Your Everyday Life

are to blame. As many as seven out of ten biologists think the world may be in the grips of a mass extinction event. It would be the sixth, as five have occurred throughout the eons of time. Sometimes, it's hard to wrap our minds around complex issues. With almost unimaginable lengths of time, it is especially hard. Think of the dinosaurs. This large group of successful creatures ruled Earth for more than 150 million years. A long time by any count—and then they vanished.[6]

Human beings are connected to the natural world in a complex web of life. Ecology isn't just for academics; every person on Earth is exposed to it every day in one form or another.

Activity: Reduce Your Garbage

Every person contributes to solid waste or trash. After it goes into the garbage, much of it is buried in landfills. Is there a way of reducing trash? Some people reduce trash by composting. Can all materials be composted?

Composting kitchen scraps is a great way to reduce the amount of trash filling up your garbage. Some communities even offer to pick up compostable materials much like they pick up recyclables.

Ecology in Your Everyday Life

Things You Will Need:

- a pencil
- paper
- masking tape
- a marker
- 2 large empty plastic or metal coffee cans with covers
- dirt to fill each can half full
- dried grass or leaves to fill each can half full
- things to be buried: a plastic spoon, a small wrapper from candy, a plastic soda bottle cap, an apple core, a banana peel, a small handful of carrot peels
- a measuring cup
- water
- gloves (if you don't like to get your hands dirty)

* **Do this during spring, summer, or fall.**

1. Have the pencil and paper ready to record observations.
2. Label one can A and one B.
3. Fill each coffee can half full. Use layers of dirt and leaves or grass.
4. In can A, put the spoon, wrapper, and bottle cap. (Spread these items out.)
5. Fill the rest of can A with layers of dirt, grass, and leaves.

Environmental Issues Affect Us All

■ **6.** Repeat steps 4 and 5 with can B and the leftover core and peels.

■ **7.** Sprinkle each container with half a cup of water.

■ **8.** Place the containers in a sunny spot outside.

■ **9.** In a week, gently spread dirt aside and observe buried items.

■ **10.** Make observations: What do the items look like? Are there any changes?

■ **11.** Put the items back into place, and cover them with dirt again.

■ **12.** If the dirt is dry, repeat step 7.

■ **13.** Repeat steps 8 through 12 three more times (a total of four times over four weeks).

■ **14.** Answer these questions: Is there a difference between the buried items? Which items appear most changed? What is happening here? How could this make a difference for landfills and trash?

As you may have guessed, can B is an example of composting. It can be done with most kitchen scraps (no meat products, such as bones or grease) to reduce solid waste.

Introduction

1. Christine Dell'Amore, "Species Extinction Happening 1,000 Times Faster Because of Humans?" *National Geographic*, May 30, 2014, https://news.nationalgeographic.com/news/2014/05/140529-conservation-science-animals-species-endangered-extinction.
2. Nigel Stork, "How Many Species of Insects and Other Terrestrial Arthropods Are There on Earth?" *Annual Review of Entomology* 63 (January 2018), abstract, PubMed.gov, https://www.ncbi.nlm.nih.gov/pubmed/28938083.

■ Chapter 1
Nature Is All Around

1. Martha Cyr, Ioannis Miaoulis, and Michael J. Padilla, *Science Explorer: Environmental Science, Teacher's Edition* (Upper Saddle River, NJ: Prentice-Hall, Inc., 2002), p. 20.
2. "Squirrels in New York City's Parks," NYC Parks, https://www.nycgovparks.org/programs/rangers/wildlife-management/ squirrels (accessed September 13, 2018).
3. Simone Wilson, "Where to Spot Wild Animals in NYC: Awesome New Seasonal Map Tracks Urban Wildlife," New York City Patch, May 19, 2017, https://patch.com/new-york/new-york-city/where-spot-wild-animals-nyc-awesome-new-seasonal-map-tracks-urban-wildlife.
4. "Wildlife Overview," City of Boulder Colorado, https://bouldercolorado.gov/osmp/wildlife-overview#insects (accessed September 14, 2018).
5. "Beaver," University of Kentucky, College of Agriculture, Food and Environment, https://forestry.ca.uky.edu/beaver_damage (accessed September 15, 2018).
6. Kristin, "About Those Acorns," Mass Audubon, October 16, 2012, https://blogs.massaudubon.org/yourgreatoutdoors/about-those-acorns.
7. "Acorn Nuttiness," Songbird Garden, https://www.songbirdgarden.com/store/info/infoview.asp?documentid=304 (accessed September 16, 2018).
8. "Cooper's Hawk Life History," Cornell Lab of Ornithology, https://www.allaboutbirds.org/guide/Coopers_Hawk/lifehistory (accessed September 16, 2018).

Chapter Notes

■ Chapter 2
Matter and Energy—It's a Big Deal

1. Martha Cyr, Ioannis Miaoulis, and Michael J. Padilla, *Science Explorer: Environmental Science, Teacher's Edition* (Upper Saddle River, NJ: Prentice-Hall, Inc., 2002), pp. 44-46.
2. Rhiannon Crain, "Habitat Feature: Milkweeds," Habitat Network, December 9, 2015, http://content.yardmap.org/learn/milkweeds.
3. "About Woodchucks," Mass Audubon, https://www.massaudubon.org/learn/nature-wildlife/mammals/woodchucks-groundhogs/about (accessed October 19, 2018).
4. "American Toad," Nature Works, http://www.nhptv.org/natureworks/americantoad.htm (accessed October 19, 2018).
5. Cyr, Miaoulis, and Padilla, pp. 45-47.
6. Kim Rutledge, et al., "Autotroph," *National Geographic*, updated January 21, 2011, https://www.nationalgeographic.org/encyclopedia/autotroph.
7. Clive A. Edwards, "Earthworms," Natural Resources Conservation Service, https://www.nrcs.usda.gov/wps/portal/nrcs/detailfull/soils/health/biology/?cid=nrcs142p2_053863 (accessed October 19, 2018).
8. "Nitrogen-Fixing Native Plants," The Pizzo Group, https://pizzogroup.com/articles/nitrogen-fixing-native-plants (accessed September 30, 2018).

■ Chapter 3
Where We Live

1. "How Does Biodiversity Loss Affect Me and Everyone Else," WWF Global, http://wwf.panda.org/our_work/biodiversity/biodiversity_and_you (accessed October 14, 2018).
2. "The History of Antibiotics," American Academy of Pediatrics, updated November 21, 2015, https://www.healthychildren.org/English/health-issues/conditions/treatments/Pages/The-History-of-Antibiotics.aspx.
3. "How Does Biodiversity Loss Affect Me and Everyone Else."
4. Neil A. Campbell, Jane B. Reece, *Biology, Seventh Edition* (San Francisco, CA: Pearson Education, Inc., 2005), p. 1211.
5. "Biodiversity," Ecological Society of America, https://www.esa.org/esa/wp-content/uploads/2012/12/biodiversity.pdf (accessed October 14, 2018).
6. "Living with Wildlife: Starlings," Washington Department of Fish and Wildlife, https://wdfw.wa.gov/living/starlings.html (accessed October 14, 2018).
7. "Sea Otter," *National Geographic*, https://www.nationalgeographic.com/animals/mammals/s/sea-otter (accessed October 14, 2018).
8. "Sea Otters: The Kelp Keystone," World Animal Foundation, http://www.worldanimalfoundation.org/articles/article/8949991/186177.htm (accessed October 14, 2018).

Ecology in Your Everyday Life

■ **Chapter 4**
Resources Are Crucial

1. "History of Solar," U.S. Department of Energy, https://www1.eere.energy.gov/solar/pdfs/solar_timeline.pdf (accessed (October 15, 2018).
2. Michael Dhar, "How Do Solar Panels Work?" LiveScience, December 6, 2017, https://www.livescience.com/41995-how-do-solar-panels-work.html.
3. Elizabeth Palermo, "How Do Wind Turbines Work?" LiveScience, April 28, 2014, https://www.livescience.com/45192-how-do-wind-turbines-work.html.
4. Jeremy Hsu, "NASA Uses Algae to Turn Sewage into Fuel," Space.com, December 16, 2009, https://www.space.com/7679-nasa-algae-turn-sewage-fuel.html.
5. Monica G. Turner, "Yellowstone Rebounded from an Epic 1988 Fire—That May Be Harder in Future," *Scientific American*, August 28, 2018, https://www.scientificamerican.com/article/yellowstone-rebounded-from-an-epic-1988-fire-that-may-be-harder-in-future.

■ **Chapter 5**
Environmental Issues Affect Us All

1. Sara Ganim and Linh Tran, "How Tap Water Became Toxic in Flint Michigan," CNN, January 13, 2016, https://www.cnn.com/2016/01/11/health/toxic-tap-water-flint-michigan/index.html.
2. "Gulf of Mexico Dead Zone," Nature Conservancy, https://www.nature.org/en-us/about-us/where-we-work/priority-landscapes/gulf-of-mexico/stories-in-the-gulf-of-mexico/gulf-of-mexico-dead-zone (accessed October 17, 2018).
3. "Water Filtering of Wetlands," National Park Service, updated April 10, 2015, https://www.nps.gov/keaq/learn/education/water-filtering-of-wetlands.htm.
4. Alina Bradford, "Deforestation: Facts, Causes & Effects," LiveScience, April 3, 2018, https://www.livescience.com/27692-deforestation.html.
5. "Success Story," National Cancer Institute, https://dtp.cancer.gov/timeline/flash/success_stories/s2_taxol.htm (accessed October 17, 2018).
6. "Dinosaur Extinction," *National Geographic*, https://www.nationalgeographic.com/science/prehistoric-world/dinosaur-extinction (accessed October 18, 2018).

Glossary

abiotic Describes nonliving parts of an ecosystem.

biodiversity The variety of species in an area.

biome A group of ecosystems that share the same climate and organisms.

biotic Describes living parts of an ecosystem.

carnivore A consumer that eats other animals.

consumer An organism that eats other organisms.

decomposer An organism that breaks down dead organisms and waste materials.

ecologist A scientist who studies ecosystems.

ecology The science of how living things interact with each other and the environment.

ecosystem All the living and nonliving things interacting in a place.

habitat A place where an organism gets everything it needs to live.

keystone species A species that plays a special role in an ecosystem.

mutualism A relationship in which different organisms benefit from one another.

omnivore A consumer that eats both plants and animals.

photosynthesis The process that producers use to make food.

population The members of one species in an area.

predation The act of killing another organism for food.

producer An organism that uses photosynthesis and makes food.

Further Reading

Books

Cunningham, Anne C. *Critical Perspectives on Fossil Fuels vs. Renewable Energy.* New York, NY: Enslow Publishing, 2017.

Herman, Gail. *What Is Climate Change?* New York, NY: Penguin Workshop, 2018.

Howell, Izzi. *Earth's Resources Geo Facts.* New York, NY: Crabtree Publishing, 2018.

Hunt, Jilly. *Saving Endangered Animals.* Chicago, IL: Heinemann InfoSearch, 2017.

Ignotofky, Rachel. *The Wondrous Workings of Planet Earth: Understanding Our World and Its Ecosystems.* New York, NY: Ten Speed Press, 2018.

Websites

NASA Global Climate Change
climate.nasa.gov
Learn more about the causes and effects of climate change, as well as the solutions.

World Biomes
kids.nceas.ucsb.edu/biomes/index.html
Dive deeper into ecology and biomes.

World Wildlife Fund
www.worldwildlife.org
Discover how you can help save animals.

Index

A
abiotic, 7
agriculture, 6
air pollution, 6, 51
alien species, 32

B
bacteria, 4, 18, 21, 23, 28
bald eagle, 7, 10
biodiversity, 28
biomass, 40
biome, 30, 32-34
biotic, 7

C
carbon, 23
carbon-oxygen cycle, 23
carnivore, 19
cell phones, 41
chemosynthesis, 21
climate, 30
climate change, 38
community, 7, 34, 38
competition, 11
compost, 6
consumer, 18, 19, 32

D
decomposer, 21, 24
deforestation, 51-53

E
Earth, 6, 23, 28, 30, 40, 42, 49, 54
ecologist, 4, 30
ecology, 4-6, 54
ecosystems, 7, 13, 17, 19, 21, 23, 30, 32, 33, 34, 44, 49, 51, 53
endangered species, 10, 47
energy, 17-24
energy cycle, 17-21
environmental issues, 47-54
    activity, 55-57
Environmental Protection Agency (EPA), 10
erosion, 6
extinction, 6, 10, 53-54

F
food chain, 19
food web, 19
fossil fuels, 38, 41-43

63

Ecology in Your Everyday Life

G
global warming, 43–44, 53

H
habitat, 10, 19, 23, 30, 32, 51

K
keystone species, 34

M
make a difference, 51
matter, 17–24
    cycles of, 21, 23–24
medicines, 4, 28, 30, 53
mushrooms, 4, 21
mutualism, 13

N
natural systems, 7, 9, 30, 38, 47
    always changing, 7, 9–10
    interactions in, 11, 13
nature, 6, 7–13, 17, 21, 23, 38, 49
    activity, 14–16
nitrogen, 23, 24
nitrogen cycle, 23–24

O
oceans, 6, 33
omnivore, 19
overhunting, 6
oxygen, 7, 23, 40, 52, 53

P
photosynthesis, 18, 21, 23, 32
population, 7, 9, 10, 11
predation, 13
producer, 18, 19, 21, 32

R
recycle, 6, 21, 49, 51
renewable energy, 38, 39–40
resources, 10, 11, 38–44
    activity, 45–46
    nonrenewable, 38
    renewable, 38, 41–44
reuse, 49, 51

S
solar energy, 40

T
trash, 49, 51

W
water, 7, 11, 18, 23, 30, 32, 33, 40, 47, 49, 51
    activity, 25–27
water cycle, 23, 53
water pollution, 47, 49
weather, 10, 11, 30, 44, 49
web of life, 28, 30, 54
where we live, 28–34
    activity, 35–37